Creation and Climate Change

By Gene Allen Groner

In Honor of the Wonderful

World Created by God

In the Beginning

To

From

Other Books and eBooks by Gene Allen Groner

Journey of a Disciple

The Garden of Eden

Native American Prayers Poems and Legends

Native American Horses

Native American Fine Art

Micah's Fine Art

Fine Art by Sassan Filsoof

Son of the Most High

These Three Remain

The Helper: a Discourse on the Holy Spirit

Hallowed Be Thy Name

Deborah: Prophetess and Warrior

Saint Teresa of Calcutta

From Shepherd to King: the Story of David

The Nature of Angels

Speak To This People: Bible Prophets

For Such a Time as This: the Story of Esther

Prayers and Poems of Christ

In the Beginning

Take Off Your Sandals, the story of Moses

The Silver-Tongued Prophet: Isaiah

My God is Yahweh: the Story of Elijah

Meditations

Evangelist Billy Graham

World's Greatest Missionary

Stairway to Heaven

2020 Poems

Full of Grace

Jesus' Hands Are Kind Hands

The Kingdom of Heaven

Genesis to Revelation: Women in the Bible

Testify

Revelation

The Cross

The Road to Emmaus

Pentecost

The Mount of Beatitudes

The Resurrection of Lazarus

Leo Tolstoy and Christ

Pride and Pollution

Introduction: I'm no expert, but I've learned some interesting things from NASA, the World Health Organization, the Centers for Disease Control, and National Geographic.

The NASA graph shown below illustrates an incredible story, one that causes me great concern, both for me and for my children and grandchildren.

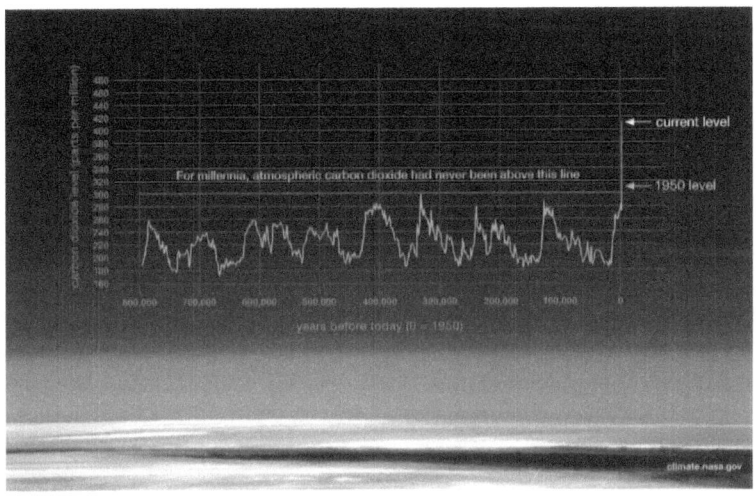

This graph, based on the comparison of atmospheric samples contained in ice cores and more recent direct measurements, provides evidence that atmospheric CO2 has increased since the Industrial Revolution, and that it had never been greater than it was in 1950 (you can see it in more detail with a small magnifying glass).

I grew up in the 1950s with a clean and healthy environment. That has all changed.

The Earth's climate has changed throughout history. Just in the last 650,000 years there have been seven cycles of glacial advance and retreat, with the abrupt end of the last ice age about 11,700 years ago marking the beginning of the modern climate era — and of human civilization. Most of these climate changes are attributed to very small variations in Earth's orbit that change the amount of solar energy our planet receives.

According to NASA, The current warming trend is of particular significance because most of it is extremely likely (greater than 95 percent probability) to be the result of human activity since the mid-20th century and proceeding at a rate that is unprecedented over decades to millennia.

For these reasons, I feel it is important for me to write this book. I am a possibility thinker, and I am hopeful for the future of the earth we live in, if we make the necessary changes in our way of thinking and in our behavior toward our planet. We need God's help—and yours.

That is my hope and my prayer.

Chapter One

The Temperature is Rising

According to NASA data, the planet's average surface temperature has risen about 1.62 degrees Fahrenheit (0.9 degrees Celsius) since the late 19th century, a change driven largely by increased carbon dioxide and other human-made emissions into the atmosphere. Most of the warming occurred in the past 35 years, with the six warmest years on record taking place since 2014. Not only was 2016 the warmest year on record, but eight of the 12 months that make up the year — from January through September, with the exception of June — were the warmest on record for those respective months.

The oceans have absorbed much of this increased heat, with the top 700 meters (about 2,300 feet) of ocean showing warming of more than 0.4 degrees Fahrenheit since 1969.

The Greenland and Antarctic ice sheets have decreased in mass. Data from NASA's Gravity Recovery and Climate Experiment show Greenland lost an average of 286 billion tons of ice per year between 1993 and 2016, while Antarctica lost about 127 billion tons of ice per year during the same time period. The rate of Antarctica ice mass loss has tripled in the last decade.

Glaciers are retreating almost everywhere around the world — including in the Alps, Himalayas, Andes, Rockies, Alaska and Africa.

The image above shows how Mount Kilimanjaro's icepack (the highest peak in Africa) is retreating over the last 50 years.

Arctic sea ice is declining, and there is less spring snowfall over the last five years.

Since the beginning of the Industrial Revolution, the acidity of surface ocean waters has increased by about 30 percent. This increase is the result of humans emitting more carbon dioxide into the atmosphere and hence more being absorbed into the oceans.

The amount of carbon dioxide absorbed by the upper layer of the oceans is increasing by about 2 billion tons per year.

Climate change is the greatest global threat to coral reef ecosystems. Scientific evidence now clearly indicates that the Earth's atmosphere and ocean are warming, and that these changes are primarily due to greenhouse gases derived from human activities.

As temperatures rise, mass coral bleaching events and infectious disease outbreaks are becoming more frequent. Additionally, carbon dioxide absorbed into the ocean from the

atmosphere has already begun to reduce calcification rates in reef-building and reef-associated organisms by altering seawater chemistry through decreases in PH. This process is called ocean acidification.

Climate change will affect coral reef ecosystems, through sea level rise, changes to the frequency and intensity of tropical storms, and altered ocean circulation patterns. When combined, all of these impacts dramatically alter ecosystem function, as well as the goods and services coral reef ecosystems provide to people around the globe, according to the National Ocean Service, a branch of the Department of Commerce.

Coral reefs are dying at an alarming rate due to climate change, killing millions of fish and plant life in our oceans and seas.

Not long ago I learned that coral is eaten by parrotfish, who then poop out the sand that is subsequently swept onto the beaches. That's where the beach's sand comes from—good, healthy coral. Parrotfish will not eat bleached out dead coral—it has no food value for them.

Making this connection with the global climate change that's the subject of this book may seem strange, but it is the scientific truth.

Thinking about beaches and sand takes me back to something I wrote a while back regarding beaches of sand. I titled this personal story Footprints in the Sand. Here's what I wrote:

Footprints in the Sand

It wasn't easy growing up without a father in my life, but it was even harder for my mother, who was faced with raising three sons by herself. I learned early on to work and help out the family—throwing newspapers, shining shoes, and selling goods door to door. In the summers I worked in the hayfields and cornfields. I became self-reliant at an early age. Since we couldn't depend on a dad to provide for us, I quickly learned to depend mainly on myself. As soon as I graduated from high school, I joined the Marine Corps. It was there that I learned about discipline, teamwork and the need to depend on others—they counted on me, and I counted on them. To survive and fulfill our mission, we needed to depend on each other.

After my tour in the Marines, I married and we soon had five wonderful children. My wife and children learned they could depend on me to take care of them and to provide a good living. It was during those early years that I first read the beautiful poem by Mary Fishback Powers, titled Footprints in the Sand. Looking back over my life I realize that I have never been alone. God has always been with me, carrying me through the difficult times when my own strength and self-reliance wasn't enough.

And more than that, I needed my family and all the other people in my life who have helped me when I needed strength and support. I owe God and everyone a debt of gratitude, and I give thanks to them all, especially in my later years when my self-reliance just isn't enough.

The simple truth is this: we all need each other—God made us that way. We need others for love and fellowship, spiritual, emotional, and material support. It is God's way, and we wouldn't want it to be any different.

In my life there have been many "footprints in the sand." Thank you, God, for a lifetime of blessings, and for my family and friends who have walked beside me on this journey of faith. I look forward to seeing them in heaven.

The poem is written on the following page.

One night I dreamed I was walking along
the beach with the Lord. Scenes from my
life flashed across the sky. In each, I
noticed footprints in the sand. Sometimes
there were two sets of footprints;
other times there was only one.

During the lowest times of my
life I could see only one set of footprints,
so I said, "Lord, you promised me,
that you would walk with me always.
Why, when I have
needed you most,
would you leave me?"

The Lord replied, "My precious child,
I love you and would never leave you.
The times when you
have seen only one set
of footprints, it was
then that I
carried you."

This poem is not only appropriate for my own
personal journey with God, it is also very
fitting for the topic of this book, Creation and
Climate Change. Climate change is causing
death and destruction. We need God's help.

Chapter Two

Storm Patterns and Infectious Disease

NASA Study: Are Supercharged Atlantic Hurricane Seasons a Case in Point?

Take hurricanes, for example. A hot topic in extreme weather research is how climate change is impacting the strength of tropical cyclones. A look at the 2019 Atlantic hurricane season provides a case in point.

After a quiet start to the 2019 season, Hurricane Dorian roared through the Atlantic in late August and early September, surprising many forecasters with its unexpected and rapid intensification. In just five days, Dorian grew from a minimal Category 1 hurricane to a Category 5 behemoth, reaching a peak intensity of 185 miles (295 kilometers) per hour when it made landfall in The Bahamas. In the process, Dorian tied an 84-year-old record for strongest landfalling Atlantic hurricane and became the fifth most intense recorded Atlantic hurricane to make landfall, as measured by its barometric pressure.

Two weeks later the remnants of Tropical Storm Imelda swamped parts of Texas under more than 40 inches (102 centimeters) of rain, enough to make it the fifth wettest recorded tropical cyclone to strike the lower 48 states. Fueled by copious moisture from a warm Gulf of Mexico, the slow-moving Imelda's torrential

rains and flooding wreaked havoc over a wide region.

Then in late September, Hurricane Lorenzo became the most northerly and easterly Category 5 storm on record in the Atlantic, even affecting the British Isles as an extratropical cyclone.

Earth's atmosphere and oceans have warmed significantly in recent decades. A warming ocean creates a perfect cauldron for brewing tempests. Hurricanes are fueled by heat in the top layers of the ocean and require sea surface temperatures (SSTs) greater than 79 degrees Fahrenheit (26 degrees Celsius) to form and thrive.

Since 1995 there have been 17 above-normal Atlantic hurricane seasons, as measured by NOAA's Accumulated Cyclone Energy (ACE) Index. ACE calculates the intensity of a hurricane season by combining the number, wind speed and duration of each tropical cyclone. That's the largest stretch of above-normal seasons on record.

So while there aren't necessarily more Atlantic hurricanes than before, those that form appear to be getting stronger, with more Category 4 and 5 events.

NASA Research Points to an Increase in Extreme Storms over Earth's Tropical Oceans

What does NASA research have to say about extreme storms? One NASA study from late 2018 supports the notion that global warming is causing the number of extreme storms to increase, at least over Earth's tropical oceans (between 30 degrees North and South of the equator).

A team led by JPL's Hartmut Aumann, AIRS project scientist from 1993 to 2012, analyzed 15 years of AIRS data, looking for correlations between average SSTs and the formation of extreme storms. They defined extreme storms as those producing at least 0.12 inches (3 millimeters) of rain per hour over a certain-sized area. They found that extreme storms formed when SSTs were hotter than 82 degrees Fahrenheit (28 degrees Celsius). The team also saw that for every 1.8 degrees Fahrenheit (1 degree Celsius) that SST increased, the number of extreme storms went up by about 21 percent. Based on current

climate model projections, the researchers concluded that extreme storms may increase 60 percent by the year 2100.

Thanks to weather satellites, scientists have identified possible correlations between the extremely cold clouds seen in thermal infrared satellite images (called deep convective clouds) and extreme storms observed on the ground under certain conditions, especially over the tropical oceans. When precipitation from these clouds hits the top of Earth's lowest atmospheric layer, the troposphere, it produces torrential rain and hail.

AIRS can't measure precipitation directly from space, but it can measure the temperature of clouds with extraordinary accuracy and stability. Its data can also be correlated with other climate variables such as SSTs, for which scientists maintain long data records.

To determine the number of extreme storms, Aumann's team plotted the number of deep convective clouds each day against measurements of sea surface temperature. They found that the number of these clouds correlated with increases in sea surface temperature.

The results of this study reflect a long line of AIRS research and three previously published papers. The researchers say large uncertainties and speculations remain regarding how extreme storms may change under future climate scenarios, including the possibility that a warming climate may result in fewer but more intense storms. But the results of this study point to an intriguing direction for further research.

New York Times Photo of Hurricane Aftermath

What Lies Ahead?

Aumann is confident future studies will reveal additional insights into how severe storms detected as individual deep convective clouds coalesce to form tropical storms and hurricanes. He notes that if you look at these clouds over the global ocean, they frequently occur in clusters.

"AIRS sees hurricanes as hundreds of these clusters," he said. "For example, it saw Hurricane Dorian as a cluster of about 150 deep convective clouds, while Hurricane Katrina contained about 500. If you look at a weather satellite image, you'll see the severe storms that make up a hurricane are not actually contiguous. In fact, they're uncannily similar to the stars within the spiral arms of a galaxy. It's one severe thunderstorm after another, each dumping a quantity of rain on the ground.

Satellite Image of Hurricane Irma

Contributing factors that increase greenhouse gases in the atmosphere include burning fossil fuels for heat and energy, producing some industrial products, raising livestock, fertilizing crops, and deforestation. Climate change leads to:

A warming ocean: causes thermal stress that contributes to coral bleaching and infectious disease.

Sea level rise: may lead to increases in sedimentation for reefs located near land-based sources of sediment. Sedimentation runoff can lead to the smothering of coral.

Changes in storm patterns: leads to stronger and more frequent storms that can cause the destruction of coral reefs.

Changes in precipitation: increased runoff of freshwater, sediment, and land-based pollutants contribute to algal blooms and cause murky water conditions that reduce light.

Altered ocean currents: leads to changes in connectivity and temperature regimes that contribute to lack of food for corals and hampers dispersal of coral larvae.

Ocean acidification (a result of increased CO_2): causes a reduction in pH levels which decreases coral growth and structural integrity.

(National Ocean Service)

Chapter Three

How Climate Change Will Affect You

A National Geographic report shows 5 ways climate change will affect you:

Torrential hurricanes, devastating droughts, crippling ice storms, and raging heat waves— all are extreme weather phenomena that can claim lives and cause untold damage. Climate change influences severe weather by causing longer droughts and higher temperatures in some regions and more intense deluges in others, say climate experts. Among the most vulnerable are communities in exposed mountain and coastal regions. In those settings worldwide, citizens are adjusting to new weather realities by strengthening warning, shelter, and protection systems.

CATASTROPHES ON THE RISE

Meteorological records show a rise in weather-related disasters since 1980. Climate change affects some weather, but experts caution against blaming it for every extreme event.

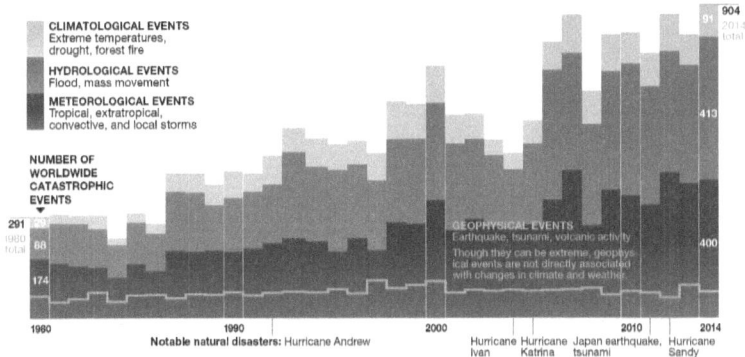

CLIMATOLOGICAL EVENTS
Extreme temperatures,
drought, forest fire

HYDROLOGICAL EVENTS
Flood, mass movement

METEOROLOGICAL EVENTS
Tropical, extratropical,
convective, and local storms

NUMBER OF
WORLDWIDE
CATASTROPHIC
EVENTS

GEOPHYSICAL EVENTS
Earthquake, tsunami, volcanic activity
Though they can be extreme, geophysical events are not directly associated
with changes in climate and weather.

Notable natural disasters: Hurricane Andrew   Hurricane Hurricane Japan earthquake, Hurricane
Ivan   Katrina   tsunami   Sandy

This chart shows the increase in catastrophic weather events since 1980, including earthquakes, tornadoes, hurricanes, fires, floods, and tsunamis.

SUCCUMBING TO HEAT

The global average temperature in May 2015 was the highest on record. In India some 2,200 people perished during a ten-day heat wave when reported highs hit 113°F (45°C). To cope, the city of Ahmadabad offered potable water and cooling centers in high-risk areas and trained health aides to treat heat-related illness.

RISING SEAS, RISING CONCERNS

Climate change may not cause a particular storm, but rising sea levels can worsen its impact. In 2012 a nine-foot storm surge from Hurricane Sandy hit New York City at high tide, making the water 14 feet higher than normal at the tip of Manhattan. Flooding destroyed neighborhoods and beaches in outer boroughs. The sea level in this area is rising by more than an inch each decade—twice as fast as the global average—and is predicted to rise 11 to 21 inches by 2050. To prepare, the city is implementing coastal resiliency measures: A multiuse project will create more green spaces for city residents as well as a system of floodwalls, berms, and retractable barriers for enhanced storm protection.

Image of a Tsunami as It Approaches Landfall

Health Risks

Climate change isn't just bad for the planet's health—it's bad for people too. Effects will vary by age, gender, geography, and socioeconomic status—and so will remedies. A recent international study in the Lancet says that many more people will be exposed to extreme weather events over the next century than previously thought—"a potentially catastrophic risk to human health" that could undo 50 years of global health gains.

Solutions are in the works. In flood-prone Benin, national health insurance has been expanded to cover diseases likeliest to increase as the world warms and sea levels rise. In the steamy Philippines, programs are helping low-income residents manage weather-related risks with loans, hygiene education, and waste and water control. Meanwhile public health experts everywhere are calling for new measures to help people stay healthy despite floods, droughts, and heat waves.

High Heat

The world will feel different in 2100, when average temperatures will have risen by several degrees. Every kind of landscape that humans inhabit will be affected: urban, suburban, rural; mountains, plains, coasts. More of the developing world will acquire life-changing modern comforts. "You'll have near-

universal saturation of air-conditioning" in warm climes by 2100, says economist Lucas Davis of the University of California, Berkeley. By powering those devices, though, we'll be contributing to global warming. If we can't find ways to turn down the heat, we'll find ways to adapt to it.

World Health Organization Report

Climate change and health

1 February 2018

Key facts

Climate change affects the social and environmental determinants of health – clean air, safe drinking water, sufficient food and secure shelter.

Between 2030 and 2050, climate change is expected to cause approximately 250 000 additional deaths per year, from malnutrition, malaria, diarrhea and heat stress.

The direct damage costs to health (i.e. excluding costs in health-determining sectors such as agriculture and water and sanitation), is estimated to be between USD 2-4 billion/year by 2030.

Areas with weak health infrastructure – mostly in developing countries – will be the least able to cope without assistance to prepare and respond.

Reducing emissions of greenhouse gases through better transport, food and energy-use choices can result in improved health, particularly through reduced air pollution.

Climate change

Over the last 50 years, human activities – particularly the burning of fossil fuels – have released sufficient quantities of carbon dioxide and other greenhouse gases to trap additional heat in the lower atmosphere and affect the global climate.

In the last 130 years, the world has warmed by approximately 0.85oC. Each of the last 3 decades has been successively warmer than any preceding decade since 1850.

Sea levels are rising, glaciers are melting and precipitation patterns are changing. Extreme weather events are becoming more intense and frequent.

What is the impact of climate change on health?

Although global warming may bring some localized benefits, such as fewer winter deaths in temperate climates and increased food production in certain areas, the overall health effects of a changing climate are overwhelmingly negative. Climate change affects many of the social and environmental determinants of health – clean air, safe drinking water, sufficient food and secure shelter.

Extreme heat

Extreme high air temperatures contribute directly to deaths from cardiovascular and respiratory disease, particularly among elderly people. In the heat wave of summer 2003 in Europe for example, more than 70 000 excess deaths were recorded.

High temperatures also raise the levels of ozone and other pollutants in the air that exacerbate cardiovascular and respiratory disease.

Pollen and other aeroallergen levels are also higher in extreme heat. These can trigger

asthma, which affects around 300 million people. Ongoing temperature increases are expected to aggravate this burden.

Natural disasters and variable rainfall patterns

Globally, the number of reported weather-related natural disasters has more than tripled since the 1960s. Every year, these disasters result in over 60 000 deaths, mainly in developing countries.

Rising sea levels and increasingly extreme weather events will destroy homes, medical facilities and other essential services. More than half of the world's population lives within 60 km of the sea. People may be forced to move, which in turn heightens the risk of a range of health effects, from mental disorders to communicable diseases.

Increasingly variable rainfall patterns are likely to affect the supply of fresh water. A lack of safe water can compromise hygiene and increase the risk of diarrheal disease, which kills over 500 000 children aged under 5 years, every year. In extreme cases, water scarcity leads to drought and famine. By the late 21st century, climate change is likely to increase

the frequency and intensity of drought at regional and global scale.

Floods and extreme precipitation are also increasing in frequency and intensity. Floods contaminate freshwater supplies, heighten the risk of water-borne diseases, and create breeding grounds for disease-carrying insects such as mosquitoes. They also cause drowning and physical injuries, damage homes and disrupt the supply of medical and health services.

Rising temperatures and variable precipitation are likely to decrease the production of staple foods in many of the poorest regions. This will increase the prevalence of malnutrition and undernutrition, which currently cause 3.1 million deaths every year.

Patterns of infection

Climatic conditions strongly affect water-borne diseases and diseases transmitted through insects, snails or other cold-blooded animals.

Changes in climate are likely to lengthen the transmission seasons of important

vector-borne diseases and to alter their geographic range. For example, climate change is projected to widen significantly the area of China where the snail-borne disease schistosomiasis occurs.

Malaria is strongly influenced by climate. Transmitted by Anopheles mosquitoes, malaria kills over 400 000 people every year – mainly children under 5 years old in certain African countries. The Aedes mosquito vector of dengue is also highly sensitive to climate conditions, and studies suggest that climate change is likely to continue to increase exposure to dengue.

Measuring the health effects

Measuring the health effects from climate change can only be very approximate.

Nevertheless, a WHO assessment, taking into account only a subset of the possible health impacts, and assuming continued economic growth and health progress, concluded that climate change is expected to cause approximately 250 000 additional deaths per year between 2030 and 2050; 38 000 due to heat exposure in elderly people, 48 000 due to diarrhea, 60 000 due to malaria, and 95 000 due to childhood undernutrition.

Who is at risk?

All populations will be affected by climate change, but some are more vulnerable than others. People living in small island developing states and other coastal regions, megacities, and mountainous and polar regions are particularly vulnerable.

Children – in particular, children living in poor countries – are among the most vulnerable to the resulting health risks and will be exposed longer to the health consequences. The health effects are also expected to be more severe for

elderly people and people with infirmities or pre-existing medical conditions.

Areas with weak health infrastructure – mostly in developing countries – will be the least able to cope without assistance to prepare and respond.

Chapter Four

World Health Organization Response

WHO response

Many policies and individual choices have the potential to reduce greenhouse gas emissions and produce major health co-benefits. For example, cleaner energy systems, and promoting the safe use of public transportation and active movement – such as cycling or walking as alternatives to using private vehicles – could reduce carbon emissions, and cut the burden of household air pollution, which causes some 4.3 million deaths per year, and ambient air pollution, which causes about 3 million deaths every year.

In 2015, the WHO Executive Board endorsed a new work plan on climate change and health. This includes:

Partnerships: to coordinate with partner agencies within the UN system, and ensure that health is properly represented in the climate change agenda.

Awareness raising: to provide and disseminate information on the threats that climate change presents to human health, and opportunities to promote health while cutting carbon emissions.

Science and evidence: to coordinate reviews of the scientific evidence on the links between climate change and health, and develop a global research agenda.

Support for implementation of the public health response to climate change: to assist countries in building capacity to reduce health vulnerability to climate change, and promote health while reducing carbon emissions.

The true costs of fossil-fuel based economies are felt in our hospitals and in our lungs. This is because the same emissions that cause climate change are also largely responsible for the polluted air we all breathe in, causing heart disease, stroke, lung cancer and infections, affecting every single organ in our bodies.

Sea level rise flooding Chennai India

Commit to invest in climate action, public health and sustainable development

"Mumbai, capital city of the Maharashtra province, is India's biggest city.

It also has some of the largest slums in the subcontinent, with large swaths of informal dwellings suffering from limited sanitation and high poverty rates. Recently, some of the streets were given a colorful makeover, which offered a stark contrast when large parts of the city were inundated by one of the worst floods the city had ever seen. The poorest neighborhoods suffer the most from these climate-induced extreme weather events, trapping them in a vicious circle of poverty and ill health."

Climate action is about people, their health and their future. By systematically including health in UN climate processes, national climate policies, as well as finance pledges, the Paris Agreement could become the strongest international health agreement of the century.

(A World Health Organization News Report)

These reports on climate change point out the need for action. Many good non-profit organizations are active in promoting improved conditions for the world's poor and disadvantaged—those who are most vulnerable to the negative effects of global climate change. There are many opportunities to volunteer and help make a difference.

As I write this, I am thinking about some of the best years of my life, as a volunteer for Habitat for Humanity International, a non-profit organization committed to providing low-income housing for the world's poor and disadvantaged.

Let me tell you a personal story about my experience in starting a new HFHI affiliate in Eastern Jackson County, Missouri.

I Had a Dream

One day I volunteered to help Habitat for Humanity in Kansas City, Missouri. I had never built a house before, but I had read about the good work they do for families in need of safe, well-built houses. Their reputation as a charitable organization is excellent. Even though I didn't have any special building skills, I knew I was a fast learner and I gave it a try. Perhaps I could carry boards or do some painting for them, and help a family in need get a good home.

I really enjoyed working with the other volunteers that day and wanted to learn more about the organization. It was right about that time that I received a letter from Jimmy and Rosalyn Carter asking for help with fund-raising and volunteering. The letter included a

return envelope and a brochure about getting involved with a local Habitat affiliate.

We didn't have a Habitat for Humanity affiliate in Blue Springs where I lived. So I wrote to their headquarters and asked for information on starting a new affiliate in my area. The material arrived, and the more I read the more I had the feeling that this was something I should organize in Blue Springs. It wouldn't be a quick and easy beginning, but I felt it was something I was being called to do.

After securing permission from Habitat headquarters to start a new affiliate, I put asked an attorney friend of mine to help me with the incorporation papers, since each affiliate is separately incorporated under the Habitat for Humanity International organization. Then I set about putting together a team of volunteers to work with me. I needed a vice-president, secretary, treasurer, and a twelve member board of directors. All of these people would also serve as builders of the houses when we were ready.

When the team was organized, and the corporate seal came in, we were ready to begin. But first, I had to raise some money for the first house. This was a problem, because I

had never done that sort of thing before—ask for donations. The only thing I knew to do was to pray about it, which I did. I got down on my knees one night and asked God to show me how to start asking for donations. Even though I didn't know how to raise the money we needed for construction, I had faith that the Lord would show me the way.

After my prayer, I turned out the lights and went to bed. I was confident that God would answer my prayer. That night, the most unusual thing happened. I had a dream. In the dream, I was seated at my desk and dialing the telephone. I called a local bank and asked to speak with the president of the bank. When he picked up the phone, I introduced myself and said I was starting a local affiliate for Habitat for Humanity, and I was calling to ask him for a donation. I had no sooner made my request, than he responded with another question. "How much do you want?" I was still dreaming, and I said, "Five hundred dollars." Without any hesitation, he asked me where he should send the check. So I gave him the address and thanked him very much, and we hung up the telephones. In my dream, everything went very smoothly, without any hesitation or delay. It was really quite amazing.

The next morning I awoke early and went to the office where I work. The dream kept coming back into my mind through the day. I was thinking, "Is that how it's done? Can it really be that easy?" So after lunch, I decided to do exactly as I had done in my dream the night before. It would either work or it wouldn't. There was nothing to lose but a phone call, and my pride if it failed. But perhaps this was God's way of answering my prayer, so I might as well give it a try. And I did. I did and said everything just like it was in the dream.

Now here's the thing. I've been in business all my life. I deal with other businessmen and I think I know how they operate. For example, you don't just phone a bank and speak with the president. He doesn't know you from Adam, and bank presidents normally don't take phone calls from strangers. They're very busy people with lots of things on their plates, and they don't have time to waste on someone they don't know. They have assistants to handle those kinds of phone calls, and if it's important the assistant will ask the bank president if he wants to take the call. More than half the time they don't. All of these thoughts were going on in my head as I considered making the phone call. Sometimes picking up the phone is like lifting weights.

I called the local bank, and I asked if I could speak with the president, just like I did in my dream the night before. And just like in the dream, the president picked up the phone and asked, "How can I help you?" I introduced myself and told him I was starting a new Habitat for Humanity affiliate in Blue Springs, and I called to ask him if he would consider making a donation. He then said, "How much to you want?" Before I could even think, I answered, "Five hundred dollars," to which he replied, "Where do I send the check?" I told him the address, and then I thanked him. Then we hung up the phones.

Wow! I could hardly believe what had just happened. The truth is, I shouldn't have been surprised. I know from past experience that God answers prayer, and sometimes he works in mysterious ways. But still, I am always amazed when it happens. This is the first time the Lord answered my prayer in a dream. I know it happened in the Bible, but I didn't know it can happen today. Now I do. God is so amazing.

Thank you Lord.

I have told this story many times, but it is always fresh in my mind. God answered my prayer. God is alive and well, and his grace and generosity are simply amazing.

President Jimmy Carter, Habitat's best-known volunteer, wrote to me asking me to get involved with HFHI. I did and I'm glad.

Habitat for Humanity is a global nonprofit housing organization working in local communities across all 50 states in the U.S. and in approximately 70 countries. Habitat's vision is of a world where everyone has a decent place to live.

Habitat works toward our vision by building strength, stability and self-reliance in partnership with families in need of decent and affordable housing. Habitat homeowners help build their own homes alongside volunteers and pay an affordable mortgage.

The Lord has said, "Men should be anxiously engaged in a good cause, and do many things of their own free will, and bring to pass much righteousness;

For the power is in them, wherein they are agents unto themselves. And inasmuch as men do good they shall in nowise lose their reward." (D.C. 58:27-28)

Chapter Five

Effects of Climate Change on Health

Centers for Disease Control – Climate Effects on Health

Climate change, together with other natural and human-made health stressors, influences human health and disease in numerous ways. Some existing health threats will intensify and new health threats will emerge. Not everyone is equally at risk. Important considerations include age, economic resources, and location.

In the U.S., public health can be affected by disruptions of physical, biological, and ecological systems, including disturbances originating here and elsewhere. The health effects of these disruptions include increased respiratory and cardiovascular disease, injuries and premature deaths related to extreme weather events, changes in the prevalence and geographical distribution of food- and water-borne illnesses and other infectious diseases, and threats to mental health.

Mental Health and Stress-Related Disorders

Mental illness is one of the major causes of suffering in the United States, and extreme weather events can affect mental health in several ways. Following disasters, mental

health problems increase, both among people with no history of mental illness, and those at risk – a phenomenon known as "common reactions to abnormal events." These reactions may be short-lived or, in some cases, long-lasting. For example, research demonstrated high levels of anxiety and post-traumatic stress disorder among people affected by Hurricane Katrina, and similar observations have followed floods and heat waves. Some evidence suggests wildfires have similar effects. All of these events are increasingly fueled by climate change. Other health consequences of intensely stressful exposures are also a concern, including pre-term birth, low birth weight, and maternal complications.

In addition, some patients with mental illness are especially susceptible to heat. Suicide rates vary with weather, rising with high temperatures, suggesting potential impacts from climate change on depression and other mental illnesses. Dementia is a risk factor for hospitalization and death during heat waves. Patients with severe mental illness, such as schizophrenia, are at risk during hot weather because their medications may interfere with temperature regulation or even directly cause hyperthermia. Additional potential mental health impacts, less well understood, include the possible distress associated with

environmental degradation and displacement and the anxiety and despair that knowledge of climate change might elicit in some people.

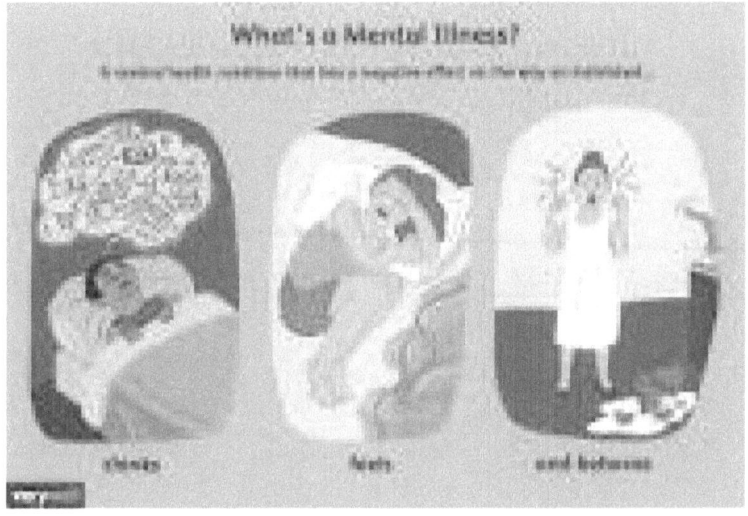

Chapter Six

Positive Things We Can Do

National Geographic Report

Ways to Curb Climate Change

Governments, businesses, and utilities are moving toward cleaner energy. But the push to trim carbon emissions begins with changing how we live.

Can one person in seven billion make a difference? Despite the furor over government reports and international conferences, climate change is a problem of personal consumption. Swiss scientists say humanity could limit the effects if each person used just 2,000 watts of power a year. The average American consumes 12,000. A Bangladeshi uses 300. The challenge is conscientious reduction in the West, writes Naomi Klein in This Changes Everything. Lifestyle choices, such as traveling less plus better regulation and technologies will help the numbers drop.

TINY HOUSE FOOTPRINT

Since 1973 the average U.S. home has ballooned by 60 percent to 2,657 square feet. A warmer world may favor a reverse trend. Jay Shafer (at left in photo), the California pioneer of living in tiny houses, built a lifestyle in 96

square feet—and helped others build pint-size homes. Developers in New York City and San Francisco have created the urban equivalent: micro-apartments.

EFFICIENT LIVING

Shrinking your space doesn't mean shrinking your life. Downsizing, experts say, can bring both psychological and financial benefits. Start by getting rid of clutter. End with lower utility bills, less space to clean, and more time outdoors.

TRANSPORTATION

If you want to use the cleanest mode of transportation, nothing beats walking or biking, which create zero greenhouse gases beyond those produced making the bike and the food you eat. From there, it's far more complicated. According to the Oak Ridge National Laboratory, transit buses use more energy per passenger-mile than cars. For long distances, you're better off flying—carpooling in the sky—or, for the ultra-prudent, taking a train.

Calculations will change as the world's fleet shifts from fossil fuels to electric. "By 2035

there will be very few conventional gasoline or diesel cars being sold," says Dan Sperling, director of UC Davis's Institute of Transportation Studies. Global trends toward mass urbanization make infrastructure planning easier. They also raise the likelihood that more people will take trains, bikes, or their own feet to get from A to B.

Eat a Healthy Diet

Using Harvard's Healthy Eating Plate as a guide, we recommend eating mostly vegetables, fruit, and whole grains, healthy fats, and healthy proteins. We suggest drinking water instead of sugary beverages, and we also address common dietary concerns such as salt and sodium, vitamins, and alcohol. It's also important to stay active and maintain a healthy weight.

The main message: Focus on diet quality

The type of carbohydrate in the diet is more important than the amount of carbohydrate in the diet, because some sources of carbohydrate—like vegetables (other than potatoes), fruits, whole grains, and beans—are healthier than others.

The Healthy Eating Plate also advises consumers to avoid sugary beverages, a major source of calories—usually with little nutritional value—in the American diet.

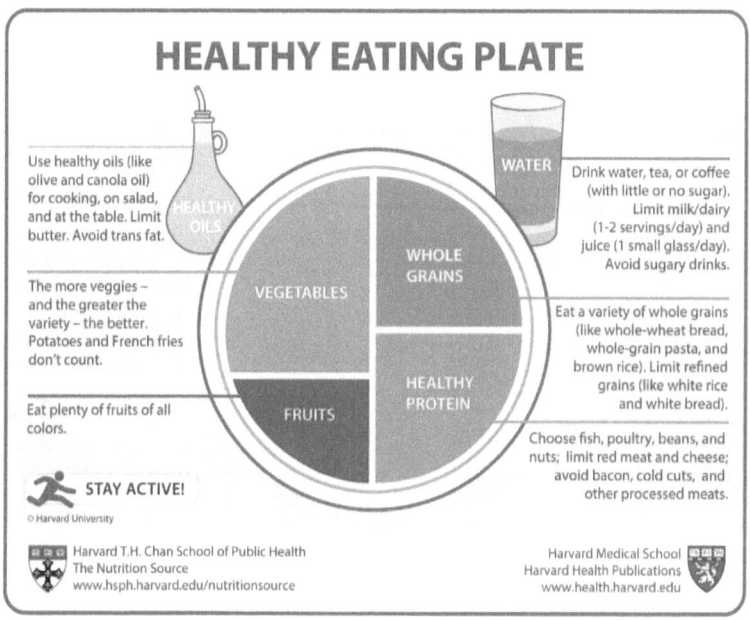

The Healthy Eating Plate encourages consumers to use healthy oils, and it does not set a maximum on the percentage of calories people should get each day from healthy sources of fat. In this way, the Healthy Eating Plate recommends the opposite of the low-fat message promoted for decades by the USDA.

The Healthy Eating Plate summarizes the best evidence-based dietary information available today. As nutrition researchers are continually discovering valuable information, The Healthy Eating Plate will be updated to reflect new findings.

Reduce Your Waste

Show kids how easy it is to reduce waste at home with these tips from Nat Geo Family.

It's not just about throwing less away. When it comes to limiting the trash that your family produces, here's how to cut back from top to bottom at home.

Reduce Your Waste

To eliminate some of the excess packaging piling up in landfills, encourage your family to shop for snacks, cereal, and pasta in the bulk section of your grocery store (if yours doesn't have any bulk items, check out a nearby natural food shop). Then, store it all in reusable glass jars.

Give back

Recycle whatever—and whenever—you can. This includes clothes you've grown out of and old toys. Instead of sending them to the dump, donate them to friends or family or to a local thrift store.

Minimize your mail

Is your mailbox always full of catalogs? Talk to your family about reducing junk mail by visiting websites like dmachoice.org and catalogchoice.org.

Reach for reusable

Instead of packing your sandwich and snacks in plastic bags, use reusable containers or cloth sacks instead. Same goes for your drink: Skip

the juice boxes and opt for a refillable water bottle.

Bag it

Reduce the amount of plastic bags clogging up our trash and oceans by shopping with reusable bags instead.

Toss in fruit

Pack an apple, a banana, or an orange. Fruit fills you up in a healthy way, plus there's no need for extra packaging. (Save the core, peels, and rinds for your compost bin.)

Nix paper napkins and wipe out paper towels

It's estimated that 17 trees are cut down for every ton of non-recycled paper. Save some branches by bringing a cloth napkin that you can wash and reuse. Then clean up in an eco-friendly way by using cloths instead of paper towels. Make your own rags by cutting up old towels or clothes headed to the donation bin.

We can make a difference. Make our planet cleaner and healthier. Don't waste. Be energy efficient. Reuse and recycle. Let's all start now.

Biography

Gene Allen Groner is a Christian writer and the author of more than 40 books and numerous articles on faith and spirituality. He lives in Independence, Missouri with his wife of 55 years, a retired public health nurse. His interests include reading and writing, gardening, and volunteer work in the community. He is listed in Who's Who in Missouri, and is a lifetime member of the National Honor Society in Psychology, Psi Chi.

Gene earned both the Bachelor and Master's Degrees with honors from Park University in Parkville, Missouri. He also attended the University of Hawaii and Saint Paul School of Theology. He and his family are lifetime members of the Colonial Hills congregation in Blue Springs, Missouri.

https://www.amazon.com/Gene-Allen-Groner/e/B077YTVSJZ

email: geneallengroner@gmail.com

Biography

Gene Allen Groner is a Christian writer and the author of more than 40 books and numerous articles on faith and spirituality. He lives in Independence, Missouri with his wife of 55 years, a retired public health nurse. His interests include reading and writing, gardening, and volunteer work in the community. He is listed in Who's Who in Missouri, and is a lifetime member of the National Honor Society in Psychology, Psi Chi.

Gene earned both the Bachelor and Master's Degrees with honors from Park University in Parkville, Missouri. He also attended the University of Hawaii and Saint Paul School of Theology. He and his family are lifetime members of the Colonial Hills congregation in Blue Springs, Missouri.

https://www.amazon.com/Gene-Allen-Groner/e/B077YTVSJZ

email: geneallengroner@gmail.com

www.ingramcontent.com/pod-product-compliance
Lightning Source LLC
Chambersburg PA
CBHW030700220526
45463CB00005B/1860